THE CELL OF LIFE
AWAKENING AND REGENERATING

Author of The God Design and Elevation

KELLY-MARIE KERR

UKCS Registration Number: 284727419

www.seekvision.co.uk

ISBN 978-1-9164137-3-3

First Printing, 2022
Printed in the United Kingdom

DISCLAIMER:
This book contains general medical information only. NOTHING in this book is intended to be a substitute for qualified, certified professional medical or psychological advice, diagnosis or treatment. You must NOT rely on the information in this book as an alternative to medical advice given by a professional healthcare provider or doctor. Consult a qualified professional healthcare provider or Medical Doctor (MD) with questions or concerns regarding practices or substances mentioned in this book that may affect your health or general wellbeing. You should always seek immediate professional medical attention if you think you are suffering from any medical condition. The medical information within this book is provided without any representations or warranties, express or implied. The medical information contained within this book is not professional medical advice and should not be treated as such. The medical information contained within this book is ONLY provided to highlight comparisons within the topics presented here, further personal research and professional guidance is always recommended.

THE CELL OF LIFE

AWAKENING AND REGENERATING

CONTENTS

PREFACE

> "There is an automatic procedure within the human body,
> which, if not interfered with will do away with all sickness,
> trouble, sorrow and death, as stated in the Bible."
>
> **Page 21, "God-Man: The Word Made Flesh" by George W Carey & Ines Eudora Perry**

The Great Regeneration is also known as the preservation of the **Sacred Secretion** and **The Threefold Enlightenment**.

It's called "threefold" because it causes changes in the body, mind, and spirit. Meaning that the benefits are felt physically, mentally, and energetically.

This study includes historical and modern perspectives so that the parallels can easily be appreciated. Let's take a proper look at the process of **The Great Regeneration** starting with THOUGHT.

THOUGHT

We don't need to look far to realise that thought is all-powerful!

> *"The moment you change your perception, is the moment you rewrite the chemistry of your body."*
> **Bruce Lipton**
>
> *"Spiritual thinking breaks open the physical cells and atoms and releases their imprisoned life, which originally came from Divine Mind (Source)."*
> **Charles Fillmore**

Water carries the imprints of our thoughts throughout the temple-body and the cells adjust accordingly, therefore our **lymphatic-water systems are key in the Great Regeneration.**

Due to this mechanism, the quality of our thoughts effects the quality of our internal system and consequently the world around us.

The body always reflects the minds choices and is fully influenced by our thoughts. In other words, the lymphatic system creates the physical outcomes of our conscious and subconscious thoughts.

This is because thoughts, emotions, and memories are vibrations that leave imprints in our lymph water.

While enjoying this study bear in mind, EVERY SEED BEGINS ITS LIFE EXCLUSIVELY IN WATER -- including ALL bodily stem cells and DNA!

By the way -- **Stem cells are the beginning of the *physical* body** and water should be considered as the medium for life. **Stem cells are another key to the Great Regeneration. The potential of stem cells is fascinating, they have the ability to self-renew and become any cell in the body.**

> "The sign of the egg represents potentiality, the seed of gener-ation (stem cell), the mystery of life."
>
> **Page 94, Dictionary of Symbols by J C Cirlot**

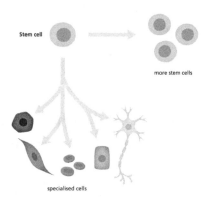

Stem cell

more stem cells

specialised cells

THE GREAT REGENERATION
– INTRODUCTION

Historical pioneers tell us that the great regeneration, sacred secretion or "divine conception" begins with the union of the solar and lunar germs in a purified body, the forces of the sun and moon in the body must be balanced in order for super consciousness to be realised.

> "The evoking of man's Solar energy can cleanse us from all these diseases; for its fire penetrates every element in our body and keeps the blood pure."
> **Page 338, Day Spring of Youth by M**

Jesus is said to have been a carpenter, *he* had to build a bridge back to super consciousness by breaking the power of the sense-mind.

We are inwardly divided, individuals; "IN DIVIDED DUALS"

The union of the lunar and solar germs is the overcoming of this inner division or "enmity" -- carnal mind vs Christ mind, black kundalini vs white kundalini, autonomic vs voluntary...

"Therefore, I say such a person, one integrated, will become full of light; but such a person, once divided will be full of darkness."
Thomas Logia 61

In the average human being, the dual power is not operating in harmony -- but once these two currents operate in harmony the regenerated being will manifest or we could say that "Jesus will resurrect".

"Jesus answered, verily, I say unto thee, Except a man be born of water and of Spirit, he cannot enter into the kingdom of God."
John 3:5 KJV

Anna Kingsford highlights this Scripture in **The Perfect Way,**

*"The man who is reborn in us is of **"Water (Lunar)"** and the **"Spirit (Solar)"** – our own regenerate Self, the Christ Jesus, and Son of Man."*

Okay, that's the historical-symbolic synopsis. Now, let's see how that translates to our actual bodies in modern scientific terms.

The esoteric lunar and solar bodies are the lymphatic (water) and respiratory (breath) systems:

- "Water" is H_2O – 2 **hydrogen** atoms and 1 **oxygen** atom
- "Spirit" is air – approx. 78% **nitrogen**, 20% **oxygen**

Thus, the "solar-spirit body" is mostly nitrogen 777, and some oxygen 888!

When nitrogen and oxygen combine in the body **nitric oxide (NO)** forms, later we'll see the INCREDIBLE role of nitric oxide in the great regeneration.

The great regeneration also involves other 'periodic' elements, the building blocks of known life. Formerly the "elements" were known as earth, water, air, and fire – more recently these elements have been shown to correspond with key periodic chemicals –

> "The water element of the ancient philosophers is the **hydrogen** of modern science; the air has become **oxygen**; the fire, **nitrogen**; the earth; **carbon.**"
> Page 292, The Secret Teachings of All Ages by Manly P Hall

These key elements make up the mass majority of our *"wonderfully made"* human body's.

Masons and other hidden groups have always known the importance of scientific elements and revered them in their art and teachings. For example, the famous Masonic number 153 signifies Hydrogen, which is atomic number 1 and has the atomic radius 53. Furthermore, 666 signifies carbon, which has 6 neutrons, 6 protons and 6 electrons – carbon is the alchemical "coal" of the ancient masters and the "earth" of material form. Carbon dioxide was alchemically known as "fixed" or "stale" air, which is interesting when you consider that breath exercises, known as pranaymas, meditations and yoga all aim to expel excess carbon dioxide, which is a poison in the body. Many mystic and religious teachers were well aware of the fact that Nitrogen, previously known as "fire", has 7 protons, 7 neutrons and 7 electrons – 777.

ALL THE MINERALS (cell-salts) IN THE BODY are formed by the precipitation of nitrogen -- the fire of life! And phosphorus -- the light of life!

The mineral body is the alchemists "inner salt" body, the solar or sun body.

This inner-invisible body is sometimes called Proteus; it is the life-maintaining principle in the body.

There are literally hundreds of examples of what we now consider periodic elements being portrayed in the allegories and metaphors of historical art and literature.

When mystics discuss the "astral" body, they are actually referring to the atomic or molecular body!

Electrons create atoms, atoms create molecules, molecules create cells, cells create tissues, tissues create organs, organs organize into systems, systems organise to create the body.

Or, in short:

Electrons > Atoms > Molecules > Cells > Tissues > Organs > Systems > The Body.

But how does the great regeneration happen and where do electrons come from? That's exactly what will be defined in this study; an organ-to-organ overview with some critical pit stops on the way.

THE "LIGHT"

Photon Light IS ELECTROMAGETIC ENERGY which takes form as both:

- **Particles (photons)**
  and
- **Waves**

In other words, a single photon IS A LIGHT PARTICLE OF ELECTROMAGNETIC ENERGY.

"God" is often described as "Light," *"God is Light, in him there is no darkness at all."* I **John** 1:5 – therefore, we could say:

> *"God is electromagnetism".*

"Light", or indeed "electromagnetism" predominantly enters the body via the nose, mouth, eyes, and fontanels.

The ears are not included because they convey ultrasonic (sound) energies which are equally important, but too much to explain in this particular book.

Photon Light is an Intelligence, an ultimate unit; inseparable, and indivisible – "God".

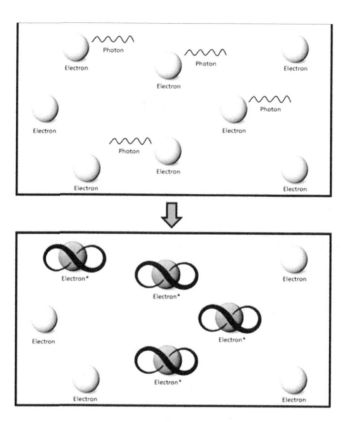

In the body Photon Light is absorbed or transformed into electrons, predominantly in the form of nitrogen and phosphorus (formers of mineral cell-salts).

The circulation and transformation of electromagnetic energy (photon light) in biological systems is the foundation of life on earth.

In the invisible, "unmanifest" realm, the sacred geometry of photons reveals the seed and flower of life.

ELECTRONS

An electron is the lightest stable subatomic particle known.

By the way: Neutrons are like electrons with zero charge and protons are like electrons with positive charge... so broadly speaking, IT'S ALL ELECTRONS.

The fundamental unit of electricity is the negative charge of the electron!

The Nobel prize laureate **Szent-Gyorgyi** famously said, *"ALL LIFE DEPENDS ON A SMALL TRICKLE OF ELECTRONS FROM THE SUN."*

The sun has been personified many times over the course of history – some examples:

- Brahma
- Krishna (the ancient Hindus call the sun Kris)
- Amun
- Abram
- Adonis / Adonai / Lord
- Mithras
- Zeus / Iesous / Jesus

Samael Aun Weor says that the *"serpentine fire (kundalini) dwells in electrons"* and referred to *"the light of the sun as being a product of the __sexual__ process of carbon."* Because… *"hydrogen 1-1 and carbon 6-6 unite to form solar fire (nitrogen 7-7) – this is the process of atoms uniting and transforming."* He further explains that *"the transmutation of carbon 12 (666) bestows upon man such human powers, that they make every technological advancement to date completely redundant."*

Electrons are what Dr Carey refers to as "gold."

> "The word **gold** comes from the word "or" or "aur" – a product of the **sun's rays** or breath of life. Life or Spirit **breathed** into man precipitates brain cells and grey matter which create or build the fluids and structure of physical man. "Or" is the seed (substance) as in W-OR-D or L-OR-D. "In the beginning was the word (seed) and the word was God", God means **power.**"
> Page 56, God Man: The Word Made Flesh by G W Carey and I E Perry

- Electrons are golden
- Electrons are a product of the sun's rays (photon light)
- Electrons are present in air/breath
- Electrons are the fundamental unit of power (electricity)

Carey also says that the receptacle of "Gold" is the Pineal Gland – *"receiver and transmitter of golden solar energy."* This has now been scientifically validated: cells in the pineal gland detect photons (electromagnetic energy) and send it throughout the nervous system (as electrons) by phototransduction.

We can further validate the electron as "gold" using Charles Fillmore's **The Twelve Powers of Man:**

> *"Through thought energy or the dynamic power of the mind, man can **release the life of the <u>electrons</u> secreted in the atoms that compose the cells of his body**... ...you weren't born to die. You were born to harness your full atomic capabilities!"*

This statement really sums up what the Great Regeneration is all about. The power stored in the millions of cells that compose the human body is infinite and accessing it is the key.

Physical science says that if the electronic energy stored in a single drop of water was suddenly released its power would demolish a six-story building! Therefore, the power stored in the millions of cells that compose the human body is infinite!

> "Particles of energy (electromagnetic photon-light) combine to form electrons; electrons group to make atoms; atoms draw together to form molecules; molecules combine and form cells; cells combine and form the different organs and substances of the body."
> **Charles Fillmore, Unity Magazine Feb 1933**

Now that we have a good understanding of the "Light" and the way it combines to form electrons and subsequently atoms such as hydrogen, carbon, nitrogen, oxygen, and phosphorus which in turn make molecules or mineral cell-salts, we can move on to discover the journey of Light through the body and how it has the potential to bring about the great regeneration.

FOLLOWING THE LIGHT (SACRED SECRETION) THROUGH THE BODY:

In the Great Regeneration, "Light" comes through the door of **brahma (fontanels)** and is received by the **claustrum,** a <u>blood/cerebrospinal fluid barrier (CSF)</u>.

FONTANEL – THE "DOOR OF BRAHMA (GOD)"

Earth has a "plasma fountain" at its north pole and, humans have a "plasma fountain" at the location of their crown chakra.

This "plasma fountain" is the fontanel in the skull -- also known as "the little fountain," or the "opening of Brahman". In early Christian mysticism, this opening was known as "Thura Iesous," -- the door of Iesous (Jesus).

Physiologically, it is a small opening on top of the skull.

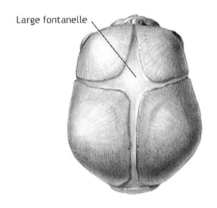

Large fontanelle

CLAUSTRUM – "BLOOD SPIRIT BARRIER"

Neurons inside the claustrum branch out and extend around the entire circumference of the brain, much like a "Crown of Thorns." A group of researchers from George Washington University found that **the claustrum can act like an on-off switch for consciousness.**

In Latin the claustrum is known as "**claustrum** hematoliquorosum". "Claustrum" means **barrier,** "hema" means **blood** and "liquor" means **spirit**; therefore, "claustrum hematoliquorosum" means "blood-to-spirit barrier."

Which is exactly true since **the claustrum is a barrier between blood and cerebrospinal fluid (CSF) -- CSF is a filtrate of blood.**

Dr Carey says, *"It is from the claustrum that the wonderful 'Christ oil' is formed."* In plain terms, the "Light" (electromagnetic energy) acting on the substances of the brain forms cerebrospinal fluid (CSF) -- "Christ Oil".

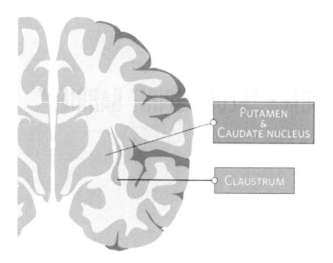

CEREBROSPINAL FLUID (CSF) – "CONVEYER OF LIGHT ENERGIES"

Cerebrospinal fluid is a filtrate of blood; **there is a perpetual exchange occurring between blood, CSF (nerve/interstitial fluid), sexual vital fluids and lymph.**

The moon regulates all of these vital fluids in our bodies – that's why the fluidic (lunar) body regenerates monthly.

At the claustrum-blood/CSF barrier, the body turns "**water** into **wine**" every second. "Water" symbolises blood and "Wine" symbolises CSF – SPIRITualised substance; CSF is saline (salty) and alkaline (electron rich).

CSF is secretly known as the *"blood of the Lamb."* The ventricles are even shaped like ram horns, hence the esoteric affiliation with Aries.

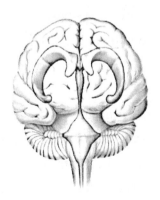

CSF IS THE MOST CONDUCTIVE FLUID IN THE
BODY, the breath (mostly nitrogen and oxygen) charges (ionizes)
CSF. The Great Regeneration relies on the charging and transmut-
ing of this subtle fluid.

> "The kundalini utilizes that which is termed spinal liquid. It
> actually ionizes CSF and changes it molecular structure and
> consequently the basic DNA structure of the entire body."
>
> **Page 243, A Beginners Guide to Creating Reality by Ramtha**

In the Metaphysical Dictionary, Charles Fillmore refers to CSF as
*"a spiritual fluid that God is continually propelling throughout man's
whole being, as the electromagnetic centre of every physically expressed
atom."*

Dr Zappaterra says, *"CSF acts as a storage field and conveyer for
light energies."*

So far, we've seen that "Light" precipitates as CSF in the claustrum, so what happens next? CSF is differentiated by the **pituitary
gland** and the **pineal gland**.

The distinct secretions and vibrational qualities of the pituitary
and pineal create two different potencies.

These potencies are known as black kundalini (pituitary) and
white kundalini (pineal) or "lunar" (fluid) and "solar" (fire/mineral)
respectively.

The pituitary and the pineal are endocrine glands.

The cerebellum controls endocrine function via the **autonomic
nervous system (ANS)** and the breath (via the vagus nerve) supplies
the power to run the entire system.

Both the sympathetic and parasympathetic nervous systems are part of the **autonomic nervous system (ANS)**. Throughout the body, parasympathetic fibres meet with sympathetic fibres to form nerve plexuses (chakras). So, the balance between these two systems improves chakra health. Parasympathetic nerves rule rest and healing, sympathetic nerves rule flight and flight – both have their uses in life and health.

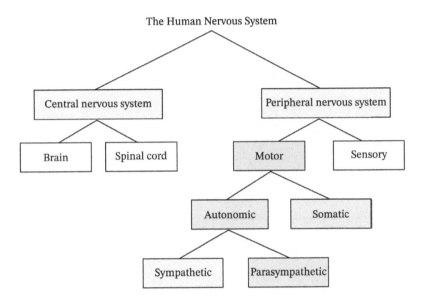

As we continue, remember: the "divine conception" or great regeneration stands for the union of the lunar and solar germs in a purified or "virgin" body.

So, let's take a look at the two potencies formed by the lunar pituitary and solar pineal.

PITUITARY GLAND – THE LUNAR POTENCY, "LUNAR GERM"

Flowing into the pituitary **CSF** becomes magnetic, "female", in its quality and action.

It's important to remember that the pituitary gland is part of the **generative (reproductive system)**, carrier of vitality (sexual energy/fluid) and the governor of the nervous systems and of the ductless glands.

The pituitary is the microcosmic moon, biblically symbolised by Mary, the mother of the Holy Child, "she" produces the **"lunar seed"** also known as:

- "Oil/Water of Life"
- "The feeling principle"
- "Soul – Fluid Body"
- "Silver" or "Milk"

But what exactly is the "lunar germ?" let's find out…

> "A lunar germ is made of matter of the **four** worlds (the light world, life world, form world and physical world)."
> **Page 464, Thinking and Destiny by H W Percival**

Mr Percival is using symbolic language to tell us that protoplasm is the "lunar germ", protoplasm is indeed made of the four key elements, in **The Secret Teachings of All Ages** Manly P Hall gives it to us more plainly,

"It is made of carbon, hydrogen, oxygen, and nitrogen... Its name is protoplasm. And it is not only the structural UNIT with which all living bodies (cells) start life, but with which they are subsequently built up..."

The **lunar germ** moves with the lunar month (13 moon cycles per 12-month solar year). It corresponds with the fluidic/water body which renews itself monthly with the lunar cycle (i.e., every 29.5 days).

Now we can move on to the second potency.

PINEAL GLAND – THE SOLAR POTENCY, "SOLAR GERM"

Flowing into the Pineal **CSF** becomes electric, "male", in its quality and action.

The pineal is the microcosmic Sun, biblically symbolised by Joseph, the father of the Holy child which produces the **"solar seed"** also known as:

- "Fire of Life"
- "The thinking principle"
- "Spirit – Fire Body"
- "Gold"
- "Honey"

But what exactly is the "solar germ?" Let's find out…

> "The electronic solar matter is the sacred fire of kundalini. When we free this energy, we enter the path of authentic Initiation." **Samael Aun Weor.**

This quote from **The Perfect Matrimony** is really useful since the same author also tells us in **Practical Magic** that… the fire of kundalini is nitrogen. Nitrogen has 7 **electrons** and as stated earlier, it forms mineral cell-salts in the body.

Dr Carey also refers to the "solar germ" as **"electrons"** and we already know that electrons combine to form atoms such as nitrogen and phosphorus, which are both formers of mineral cell-salts.

The solar germ corresponds with the fiery/electric mineral body which renews itself yearly with the sun (i.e., on your birthday).

> THE SOLAR GERM IS ELECTRONS PREDOMINANTLY TAKING FORM AS NITROGEN AND PHOSPHORUS ATOMS, WHICH SUBSEQUENTLY FORGE MINERALS (CELL-SALTS).

QUICK RECAP:

Thus far we have seen: Light enters through the fontanel, contacts the claustrum and precipitates into CSF (oil), which is then differentiated into two streams by the pituitary and pineal. The two potencies are the "lunar germ, protoplasm, pituitary stream" and a "solar germ, electron mineral, pineal stream."

So, what happens next? Well, according to Dr Carey the CSF then,

> "...*flows down into the red nucleus and consequently the corpora quadrigemina **(olivary bodies)**, into the descending tract known as the rubrospinal tract, through the reticular formation in the pons and medulla to **the lateral column of the spinal cord.**"*

So, let's take a brief look at these body parts.

RED NUCLEUS – "RED LOTUS OF SHAKTI"

The red nuclei monitor the function of the **cerebellum** and are a relay station for spinal nerve tracts. They have input connections from the cerebellum, substantia nigra and hypothalamus and output connections with the corpora quadrigemina (olivary bodies) and the three rubro nerve tracts.

There are two red nuclei, one on either side of the spinal cord. Their output pathways **cross over** immediately after leaving each red nucleus, **this is the crossing of the Ida and Pingala nadis.** The Ida and Pingala nadis are the left and right sides of the autonomic nervous system (ANS) -- **including** sympathetic *and* parasympathetic nerve fibres.

> The diagonal cross of Ida and Pingala is actually part of the **double cross** (union jack), that many esoteric anatomists write about it. *"The pineal and pituitary body secrete the positive and negative substance along **nerves that cross in the medulla."***
> **Page 157, God Man: The Word Made Flesh.**

The cross formed by Ida is Pingala is actually part of a **double cross** which forms the site of the "crucifixion" as we will see later.

The other part of the **double cross** is formed by the vagus nerve, this will be shown later.

> "Golgotha is the base of the human skull, where the spinal cord enters the brain. At this point occurs a **double nerve crossing**, made by Ida and Pingala and the vagus nerve."
> **Page 144, God Man: The Word Made Flesh.**

In **Vallalar's Vision of Nuclear Physics**, T Thulasiram suggests that the red nucleus gave rise to the shakti yoga symbol of the red lotus.

OLIVARY BODIES – "GETHSEMANE", "THE OLIVE GARDEN"

Dr Carey refers to the olivary bodies (corpora quadrigemina) as **"the lower part of the pineal"**. The olivary bodies are biblically symbolised by "Gethsemane", "a place near the medulla oblongata, **with olives on either side."**

> *"The olives contain the oil; they are the reservoirs – the relay stations"*
> **Page 110, God Man: The Word Made Flesh.**

The olives are a collection of brainstem nuclei, the inferior olivary complex **functions as a relay station between the spine and cerebellum.**

LATERAL COLUMN OF THE SPINAL CORD

After travelling through the red nucleus, olivary bodies and crossing sides in the rubrospinal tract the two potencies come to **the lateral columns of the spinal cord** and subsequently supply the autonomic nervous system.

At the rubrospinal cross, the pituitary potency crosses from the left into the right side of the autonomic nervous system and the pineal potency crosses from the right into the left side of the autonomic nervous system.

This coincides with Harold Percival's description in **Thinking and Destiny,**

> *"From the pituitary the "***Lunar Germ***" descends through the nerve plexuses on the right side of the autonomic nervous system along the digestive tract."*

He then says, "*the lunar germ takes **one week** to reach **the solar plexus and keeps descending.**"*

The process that occurs in the solar plexus and specifically the spleen is of particular intrigue, but first the potencies must travel through the semilunar ganglion to get to the solar plexus. So, we will visit the semilunar ganglion next.

SEMILUNAR – "THE SEA OF GALILEE"

> "...the two nadis (Ida and Pingala) **converge into the body through the semilunar ganglion, where they merge into the solar plexus.**"
> **Page 48, GOD MAN: The Word Made Flesh by G W Carey**

Carey links the semilunar ganglion to "<u>GENE</u>sareth" which the metaphysical dictionary claims to be, **the sea of divine life** and the **garden of riches.**

The semilunar ganglion is:

- Two sympathetic ganglions, one on each side of the solar plexus
- Comprised of several smaller ganglions, which surround the solar plexus
- Entwined with the **vagus nerve**

Together the solar plexus and semilunar are known as *"the grand centre of all the ganglions and plexuses of organic life."*

> *"The semilunar ganglion consists of many lesser ganglions matted together in a glandular like shape."*
>
> **Vol. 1, The anatomy of the human body, by C. Bell, J Bell 1816**

Having passed through the semilunar ganglion the two potencies arrive in the solar plexus, which we will now take a look at.

SOLAR PLEXUS – "BETHLEHEM"

> "Every 29.5 days a seed is born in, or out of the **solar plexus** –
> the oil (lunar) unites with the mineral salts (solar — remember
> nitrogen and phosphorus form minerals) and thus produces
> the monthly seed which goes into the vagus."
> **Page 90, GOD MAN: The Word Made Flesh by G W Carey**

In other words, three streams unite in the solar plexus to form the
monthly seed:

1. The pineal stream (electric)
2. The pituitary stream (magnetic)
3. The breath (via the vagus)

Nerves from the solar plexus extend down to the reproductive
(sexual) organs of generation, which is perfect since Percival and
Carey both say that a portion of the seed continues to travel down-
ward after its conception in the solar plexus.

In the spleen we find the *actual* modern scientific parallel of
this historical symbolic description, so let's head over there now.

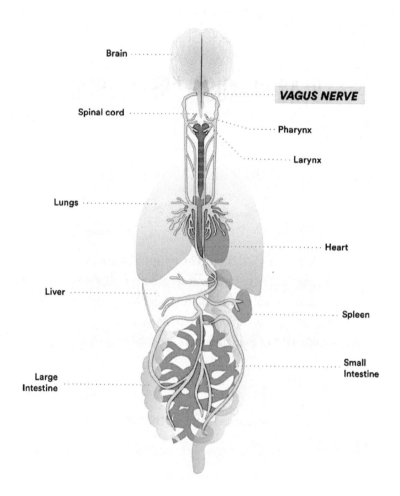

Brain

Spinal cord

VAGUS NERVE

Pharynx

Larynx

Lungs

Heart

Liver

Spleen

Small
Intestine

Large
Intestine

SPLEEN – "THE MANGER" –
BIRTHPLACE OF "THE JESUS SEED"

In this section we finally get to see what the actual "Jesus" seed is under modern scientific terms! But before this exciting key is revealed it's important to gain a little basic knowledge about the spleen.

The spleen is part of the lymphatic-water system.

The waters of the lymph system are very high in sodium, as are CSF and blood. The spleen is considered the "sodium organ." Sodium and other mineral cell-salts passing in and out of cells create the bodies electricity.

> "Not only does this electrical system run the body, _but it is actually the subtle substance with **which thinking is done.**_"
> **Page 66, The Zodiac, and the Salts of Salvation by G W Carey**

In other words, the popular manifestation mantra "thoughts are things" is completely true! **Thoughts have form** -- MINeral or "MINDeral" cell-salt form! So, the imprints made by thought in the waters of the body are of sodium and other mineral cell-salt essences. Thought is literally creating the body with every breath.

Now, here's the moment you've been waiting for... A full disclosure of the "seed" that is formed by the lunar and solar potencies!

Please continue to bear in mind that -- *there is a perpetual exchange occurring between blood, CSF (nerve/interstitial fluid), sexual vital fluids and lymph.*

The spleen contains the all-important **GERMINAL CENTRE**, symbolically known as the "manger", this is the precise place where "Jesus" is born. *"Jesus is a germ of life"* **G W Carey**

> "...The spleen mysteriously creates cells, it does this by "the enclosing of a minute body from the cerebrum within a case (Moses' basket). Thus, within the spleen is formed the TRUE PHYSIOLOGICAL SEED CELLS OF THE BODY..."
> **Page 111, Zodiac, and the Salts of Salvation by G W Carey**

Germinal centres are sites of multipotent stem cell and white blood cell production.

Hilton Hotema tells us, that in their infancy the <u>multipotent stem cells</u> produced in the lymphatic spleen are no different to procreative (sexual) cells.

> "...up to that moment, the life and conduct of the male and female gametes (infant sperm and ovum cells) present nothing different from that of the lymph cells."
> **Page 4, The Facts of Nutrition, Hilton Hotema**

The lymphatic system, including the spleen transports lymph, a fluid containing white blood cells, throughout the body. **According to Dr Carey the great regeneration invigorates white blood cell presence and production, thus improving the vitality of the body along all lines.**

> **"The process of regeneration causes the white cells of the blood to overcome the prevalence of red cells.** *Therefore, the flesh becomes transparent, and he manifests more and more of the Father, he is no longer man but has become a God."*
> **Page 171, God Man: The Word Made Flesh by G W Carey**

Multipotent stem cells are infant cells which have the ability to become many types of bodily cells.

In the centre of stem cells lies the solar potency manifested as **nucleoli, which are rich in... yes, you guessed it NITROGEN AND PHOSPHORUS!**

The image shows the radiance of nitrogen and phosphorus in stem cell nucleoli.

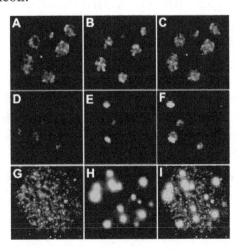

Anna Kingsford describes this process beautifully on page 144 of The Perfect Way,

> *"In her bosom (protoplasmic lunar-body), **is conceived the bright and holy Light, the Nucleoli.**"*

In other words, the "SOLAR GERM" – Electric, pineal potency (d below) unites with the "LUNAR GERM" -- Protoplasm, pituitary potency (b below), and the breath (supplied by the vagus) to form stem cells – "Jesus".

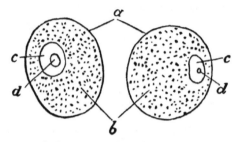

Cells of Round or Oval Form.
Border of the cell or cell-wall; *b*, cell substanc
c c, nuclei; *d d*, nucleoli.

Anna Kingsford goes further to say that the infamous "white stone" of philosophers is the solar nucleoli of stem cells,

> *"The white stone has always been the object of special veneration, it is the well-known symbol of the Divine Spirit, **the nucleoli of the Cell, the Sun of the system,** the Head of the Pyramid."*
>
> **Page 147, The Perfect Way by Anna Kingsford**

Professor Hotema explains this process most poetically,

> *"The Spirit, a fiery nucleus of noetic intelligence, is plunged into the fluidic habitat of a body of watery flesh."*
> **Page 48, The Son of Perfection (Part 1) by H Hotema**

In summary, Solar Nucleoli + Lunar Protoplasm + breath = Stem Cell

On page 21 of, **The Facts of Nutrition** Professor Hotema says,

> *"The cytoplasm (protoplasm) of the cell surrounding the nucleus is the negative (lunar-alkaline) element. The nucleus is the positive (solar-acid) element. THIS MAKES THE CELL A BIPOLAR MECHANISM and the acid-alkali balance is imperative for life."*

He is telling us that EVERY cell is bipolar (lunisolar) featuring the polarities of attraction and repulsion.

In his book, **Secret of Life**, George Lakhovsky stated that the body cell is composed of its nucleus containing chromosomes (acid), surrounded by cytoplasm (alkaline) and a cell wall. He said that the cells are bipolar mechanisms, and that chromosomes are tiny radio antennae, which pick up and receive solar rays and convert them into electric currents in accordance with the law of animation.

The production of the **physiological stem cells** of the body are what the masters refer to as "Jesus" being born in Bethlehem, the solar plexus which innervates the spleen.

Next, the seeds journey descends through the vagus to the procreative organs. In the procreative region we'll see how the seed is further transmuted or transformed.

> "The germs of life, take on human form as they enter the stomach and spleen. Then, at the second stage, these human cells are taken down (fall) by Saturn into Tartarus (hell). Meaning that in the organs of procreation they are conjoined with "animal germs" (procreative "goat" germs).
>
> **Page 285, The Zodiac, and the Salts of Salvation by G W Carey**

The seed is "taken down to hell" via T12 as we will now see.

THORACIC VERTEBRAE 12
(T12) – "GILGAL"

T12 is the place where the **solar plexus** meets the **spinal cord** via the **semilunar ganglion,** it is also the point where the spinal cord tapers off into "Sodom and Gomorrah."

At this junction, the higher forces raise a portion of the seed up the spinal cord by what Percival calls "automatic reclamation", and the lower forces pull the seed down to the "fish gate" which leads to the genitals, where there's an opportunity for "voluntary reclamation."

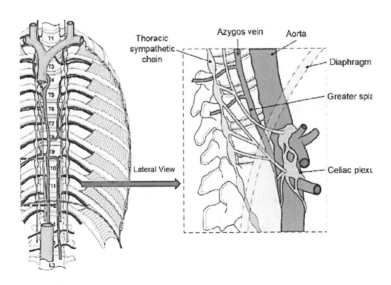

Biblically, this location is known as **"Gilgal" -- a very significant place:**

- Gilgal is the place where Saul was made king over Israel
- Gilgal is the place where Elisha neutralized poison
- Gilgal is the place where Elijah was taken up in a chariot of Fire

"Gilgal: A circle or rolling away; the place where the 12 stones (12 thoracic vertebrae) were set up, the place of the "pass over." **Refers in anatomy to the 12th thoracic vertebra, at which place the semilunar ganglion connects.** At this point the seed or ark enters Jordan or the spinal cord."
Page 17, God Man: The Word Made Flesh by G W Carey

- T12 marks the **centre** of the human organism
- T12 the centre of the star or wheel of life
- T12 is analogous with the "Hrit Padma" where "God" is said to be located.

"The Hrit Padma or "Sacred Heart" is located three finger-widths below the heart and is considered in Vedic teachings to be **the seat of the soul.**"
Pg. 46, Healing Mantras by Thomas Ashley-Farrand

The metaphysical dictionary tells us that Gilgal means "whirlwind." The "whirlwind" is the DNA double helix formed by the sacred geometry of the ratcheting dodecahedron in the pre-existent invisible realm.

This is why the dodecahedron is considered "the ascension vehicle," DNA and its dodecahedral scaffolding in the light world is the master of transformation. As proven by epigenetics our thoughts can rewrite our genetic DNA codes.

The DNA double helix looks like two intwined serPENTS winding up as the dodecahedron turns. Each side of the dodecahedral structure is a PENTagon.

The formation of DNA is often described as a spiral staircase or ladder:

- The vertical sides of the DNA "ladder" are essentially **phosphorus**
- The horizontal "rungs" of the DNA "ladder" are essentially **nitrogen**

We'll learn more about nitrogen and phosphorus and their role in the great regeneration later, but basically the preserved and reabsorbed procreative essences go towards DNA production. Not to mention DNA resides in the nucleus of every stem cell, along with the nucleoli.

> Since DNA is the very fabric, or programme for life it's as though the ancient masters believed that our bodies are animated from this central point (T12).

Let's take a pitstop on the journey to unpack some key facts about DNA as these are fundamental to understanding the great regeneration.

DNA

Under a microscope the cross section of DNA reveals the form of the seed of life.

In the book of Revelation, chapter 19:13-14 "God" is described as being *"clothed in a fabric dipped in blood"*. Fabric symbolises DNA Chromosomes, so in other words Scripture is telling us that – **"God" or Light is IN DNA.** The fabric is *"dipped in blood"* because DNA is present in literally in every blood cell of the body.

> Thus, *"the life of the flesh"* (DNA), truly is *"in the blood"*
> **Leviticus 17:11**

The cells of all living beings shine their light (electromagnetic energy), and DNA is the source of this light. When the photon-light radiated from DNA is absorbed by the body, it forms nitric oxide. Nitric oxide is key to the great regeneration, as we will see later.

> *"DNA is an aperiodic crystal that traps and transports electrons with efficiency **and that emits photon light** (electromagnetic waves)."*
> **Page 109, The Cosmic Serpent by Jeremy Narby**

The Secret Doctrine of the Rosicrucian's by Magus Incognito states that the alchemist's "mercury" is "white nitrogen" or "the secret fire." I wonder if our modern nitric oxide is the historical "white nitrogen."

Nitric oxide stimulates the synthesis of ATP which is essential for ALL CELLULAR REGENERATION.

DNA is the master of transformation!

DNA is the molecule of life! DNA is the same for all species -- the genetic information of roses, bacteria, animals, and humans is all being coded by the <u>nitrogen</u> bases in DNA -- Cytosine, Thymine, Adenine, and Guanine. **This is the universal language of 4 letters (CTAG).**

Adenine

Guanine

Thymine

Cytosine

EVERY CELL IN THE WORLD CONTAINS DNA -- and **every cell is filled with salt water.** In many cultures it is a spiritual belief that salty ocean water is the essence of all life. The concentration of salt inside cells is similar to that of the worldwide ocean. DNA is literally bathed in "seawater", which plays an integral part in establishing DNA's double helix structure. DNA's four nitrogen bases are insoluble in water so they,

> *"...tuck themselves into the centre of the molecule where they associate in pairs to form the rungs of the ladder; then they twist up into a spiralled stack to avoid contract with the surrounding water molecules. **DNA's twisted ladder shape is a direct consequence of the cell's watery environment.** DNA*

> *goes together with water, just like mythical serpents do."*
> **Jeremy Narby, The Cosmic Serpent**

Scientists have discovered that H_2O forms a "spine of hydration" around our DNA. This means that it has a "shell" of water around it, helping our DNA to operate and ensuring that everything in our body works smoothly. This hydration shell mimics the unique spiral shape of DNA.

When DNA replicates, its helical structure partially uncoils. During this uncoiling, DNA separates when complementary nitrogen bases are picked up from the cytoplasm of the cell. This process continues until the original DNA molecule has successfully duplicated itself, then cell division begins.

Dr Carey describes the process of DNA formation like this:

> "When Divine Word speaks the mineral atoms of its body into a certain combination, a germ, which is the nucleus of the form to be manifested materializes. This little plexus (germ) of intelligent atoms then commences to attract to its centre by the law of chemical affinity, other atoms known as oxygen, hydrogen, nitrogen etc., and thus materializes them, until the building is completed according to the plan of the architect or designer."
> **Page 53, The Wonders of the Human Body, George W Carey**

Before we get back to the organ-by-organ account of the great regeneration we simply must consider the absolute magic of water.

WATER AND THE LYMPHATIC SYSTEM

Water emits light (photons -- electromagnetic energy), it is an electromagnetic dipole, meaning that it's both positively charged and negatively charged:

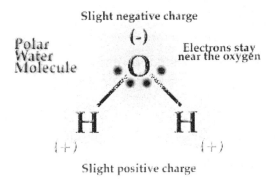

Therefore, **lymphatic (water) system is also electromagnetic – it is our light-energy body!**

- The lymph system is called the White Blood System in many cultures.
- The Lymph System is the gateway for communication between blood, CSF (nerve/interstitial fluid), sexual vital fluids and lymph fluid.

- The lymph system is the oxygen delivery and purification system at the cellular level.

Everything in the body is made of cells. All generation and degeneration of the body happens at the cellular level -- **The Lymphatic System is electro-magnetic and maintains the health of the entire cellular terrain.** Its movement is **upward, towards the neck, against the force of gravity.** Our bodies are 72% water, and the lymphatic system is the most extensive system in the body.

When the lymph system is compromised there will be a change in the bodies pH. **pH stands for potential Hydrogen, which basically refers to the number of electrons available.**

Water transforms according to the vibrational frequencies emitted in its atmosphere. Gratitude, love, and musical harmonies improve the molecular structure of water, whereas negative, violent, or hateful emotions and sounds distort it.

As stated earlier, literally EVERYTHING that we are exposed to creates imprints on our lymph-water body.

The laws of the universe are written in water. "The Golden Ratio" is found in molecules melt water. In regular water, the angle between hydrogen atoms varies from 104 to 107 degrees, IN MELT WATER THIS ANGLE IS 108 DEGREES AND THE RATIO OF THE LENGTH OF HYDROGEN TIES 0.618. This is a special state of water, in melting from its frozen state, water deletes all information from its memory, retaining only one program – **the program for life.**

Remember, many cultures believe that water is the essence of all life, even the Bible book of Genesis neglects to mention the creation of water… in fact, this most popular account of creation insinuates that water is pre-existent.

Everything in nature's perfection has been created in accordance with **the program for life**. From the DNA molecule to the formation of leaves and petals and the intricate detail of the spider spinning its web!

The "creator" of this world transmitted the **program of life** TO EVERY LIVING CREATURE THROUGH WATER.

This is why the lymphatic system and our body's pH level (acid to alkaline) is so important!

- What is acid? Acid is devoid of free electrons.
- What is alkali? Alkali is FULL OF ELECTRONS.

So, our body should be slightly alkaline. Electrons are teeny tiny magnets holding the cells of the body together, maintaining life and slowing degeneration.

Having seen the power of water as literally the catalyst for life and health we can get back to the organ-by-organ account of the great regeneration.

We can't visit the procreative organs without first mentioning the process that occurs in the intestine, thus we'll journey there next.

INTESTINE – "ALCHEMIST'S ALEMBIC"

The true work of Alchemy is the regeneration of the physical body.

The "alchemist's alembic" (bottle) is the "vas hermetic" (hermetic vase), fashioned after the lower intestine of the human body.

The stomach and small intestines are the vessels in which the sustaining minerals of life are prepared (these support the "birth of the seed" which occurs in the spleen). Without digestion the great regeneration is physically impossible.

Digestion serves many bodily functions:

- It renovates blood
- It renovates all the organs of the body
- It renovates the nerves and the brain
- It stimulates the activity of **nitric oxide (NO)**.

AND MOST IMPORTANTLY,

- It sets the vital mineral cell-salts contained in food free, creating a mineral base for the body

The mineral base is carried into circulation throughout the rest of the body via absorption into the blood and lymph at the small intestines.

Taoist master, Mantak Chia says, *"The Structure of the Intestines is like a battery – the stomach is where we store our energy. The "alembic" is the last twelve inches of the lower intestine -- it must be sealed, so none of the vapours can escape from the alembic as it does his work."*

This is part of the purification process; intestinal fermentation produces methane gas (CH_4) -- one carbon atom and **four hydrogen atoms. Hydrogen is the basis for all that exists! The VERY beginning of the new cells (self) or body.**

Speaking on digestion, Dr Carey warns that the seed can be ruined by, *"alcoholic drinks, or gluttony that cause ferment-acid and even alcohol in intestinal tract—thus "No drunkard can inherit the Kingdom of Heaven."*

Historically, the essence produced by digestion in the small intestine was known as "First Matter."

First Matter (Virgins Milk):

"First matter" is also known as "Lac Virginis", "Virgin's Milk" and "Prima Materia".

It is described as an "oily water" or "fatty lymph". It supplies the blood with the energy derived from food (mainly hydrogen).

First Matter is comprised of the common elements – Carbon, Hydrogen, Nitrogen, Oxygen, and Phosphorus; all the constituents of DNA.

*"Both our DNA and our RNA are composed of carbon, hydrogen, oxygen, and nitrogen, as well as phosphorus – **nucleic acids are rich in phosphorus**...*

> ...Along with Earth, Air, Water, and Fire, the master alchemist recognized a fifth Element, a sort of philosophic secret, and this Element was that of Spirit...
>
> ...The alchemists believed that light represented "Spirit", and as the alchemist Dennis Hauck writes, '**Since phosphorous absorbs light and even glows with it, it is the agent in the body that absorbs the spiritual light.'"**
>
> **Page 121, Esoteric Science Vol 1. J.S Williams**

In other words, alchemists believed that phosphorus was the "fifth element" – "Light" or "*Spirit*"

This may seem to contradict the earlier notion of air (nitrogen and oxygen) being "spirit", so it's important to understand that phosphorus is part of the nitrogen family, phosphorus is alchemically considered to be, "light" and nitrogen is considered to be "plasma" or "fire."

In **The Pre-existent Man,** Professor Hotema explains that Nitrogen (fire) and Phosphorus (light) produce life – which is true in the sense that they are formers of mineral cell-salts and are fundamental parts of DNA.

DNA IS QUITE LITERALLY THE "QUINTESSENCE" of life! "Quint" means five, so DNA is the "five-essence" comprised of the five fundamental elements:

Elemental	State	Periodic Element
Earth	Solid	Carbon
Water	Fluid	Hydrogen
Air	Gas	Oxygen
Fire	Plasma	Nitrogen
Spirit	**Light**	**Phosphorus**

The last step of digestion is that the the small intestine absorbs the molecules (fuel) formed from digested food. These molecules can be considered "Prima Materia", the first matter from which all cells are produced and develop.

The small intestine also plays a role in phosphorus absorption. So, now would be a good time to shed a little light on phosphorus which was renamed "lucifer" in the Latin Bible.

PHOSPHORUS AND CARBON – "The Sides of the DNA Ladder"

> "...the moral of this gospel of the flesh, is to produce plenty of phosphorus by means of good eating and drinking. Those who say, "Let us eat and drink, for tomorrow we die" are diametrically opposed to the Holy Scriptures."
>
> **Page 147, Modern Doubt and Christian Belief, A Series of Apologetic Lectures Addressed to Earnest Seekers After Truth by T Christlieb**

As explained at the beginning of this study, "Chi" or "Life Force" is essentially "photon light" (electromagnetic energy).

When electromagnetic energy is absorbed by the body it produces nitrogen and phosphorus. The **"seed-electrons"** or "**cell-nucleoli**" (**solar germs**) that combine with the protoplasmic lunar germ contain phosphorus and nitrogen.

The biological (in the body) form of phosphorus is called phosphate, which basically consists of phosphorus and oxygen.

Phosphates improve nerve function and nutrition, are active in creating the bone matrix, assist muscle tissue and are necessary for sexual function and reproduction. Phosphates are also considered vital for energy production and storage as well as activating hormones, repairing tissue, producing DNA and RNA, and it has a critical function in stem cell development, proliferation, and differentiation.

When the bodies pH level becomes too acidic phosphorus availability decreases.

Phosphates produce sugars such as "deoxyribose," **the sugar-phosphate back bone of DNA.** Deoxyribose is a "PENTose sugar" -- meaning a **five-carbon** sugar.

This five-**carbon** sugar with a pentagon structure, forms the vertical sides of the DNA ladder. Looking more closely at the word **car**bon affords us the opportunity to make some interesting links… for example, Dr Carey calls the seed, "*Osiris in his **CAR***". And Strong's Concordance sites "car" as being from the origin "**karar**" meaning whirling, dancing, or HELIX…

"Karar" is a derivative of "**karyo**" meaning nucleus or seed! And funnily enough "**karyotin**" **means the same as** "**chromatin**"**which is just DNA and protein IN THE CELL NUCLEUS.**

Having taken an important diversion to appreciate the importance of water and the molecules of DNA we will now return to the organ-to-organ analysis of the great regeneration, but this point on the journey the seed has arrived at the procreative organs.

PROCREATIVE ORGANS – "SODOM AND GOMORRAH"

According to Percival, *"the lunar germ takes one week to reach the solar plexus and keeps descending... ...After the second week, the lunar germ is said to reach its lowest point in the large intestine where it crosses over to the left side via the coccygeal plexus in front of the coccyx."*
Page 964, Thinking and Destiny by H W Percival

"The early Jews called the sacral and coccygeal plexuses the cities of Sodom and Gomorrah."
Page 22, The Occult Anatomy of Man by Manly P Hall

The <u>sacral</u> and <u>coccygeal</u> (root) plexuses innervate the procreative organs. The organs of this lower region **work together** to process 1. The procreative *seed* (sperm or ovum) and 2. The procreative (sexual) vital *fluid*.

Therefore, this section includes:

1. The gonads (testes and ovaries) – which process procreational *seeds*

2. The prostate – which processes seminal *fluid*

> *"The semen is excreted by the Prostate as stated, and the Zoa by the Gonads. The life Essence of the Body."*
> **Page 121, The Son of Perfection by H Hotema**

These two, the *seeds* and the *fluids* don't combine unless ejaculation occurs.

The time which Life Force energy spends in the organs of procreation is referred to as "hell" because of the torment induced by sexual temptation while the long task of **processing sexual vital fluid** occurs.

Dr Carey tells us,

> *"The blood secretes plasma into the prostate."*
> **Page 78, God Man: The Word Made Flesh by G W Carey**

So, it's time to take another quick pit stop to appreciate the role of blood plasma.

BLOOD PLASMA – The Golden "Fluid" Portion of Blood

Plasma is what's left when all the blood cells (red, white and platelets) are separated from whole blood. The remaining golden coloured fluid **is 90–92% water!**

Plasma contains a significant amount of nitrogen, plus critical mineral cell-salts necessary for sustaining health and life.

These cell-salts are also constituents of the procreative products (seed and fluid). In fact, Samael Aun Weor affirms that all 12 "cell-salts" are contained in the sexual essences.

In plasma, electrons are ripped away from their nuclei, forming an electron "sea". This gives it the ability to conduct electricity.

In other words, electrons in plasma ARE FREE -- it is free electrons that create and maintain life, as well as establishing all energy happening at every level in the body.

All mineral cell-salts are electron rich, so when science talks about the benefits of minerals it's due to their concentration of electrons. **In essence, a key to understanding nutrition is electron flow.** Live foods have the most electrons of any food. Alkaline water has biologically active natural hydrogen and is therefore electron rich.

> *"Due to its interaction with electromagnetism, plasmas tend to form into cellular and filamentary structures".*
> **Electric Universe Magazine**

Examples of filamentary plasma structures include veins, arteries and nerves, tree branches, cracks in rocky surfaces, lightning, electric sparks, the sun etc.

Plasma is known as the "fire state" -- The "fire" in the body is the plasma in the body.

> *"...blood plasma, is the saline solution (salt water) which vitalizes the blood cells swimming in it. This plasma is the salt sea of the body in which live the physiological FISH; for the CELLS ARE FISH IN THE HUMAN SEA."*
> **Page 288, The Zodiac and The Salts of Salvation by G W Carey**

Returning our focus to the procreative organs and sexual essences, we find that teachers from various cultures across the world, refer their preservation in different ways. "Brahmacharya" is one guise.

Guru Rajneesh of Rishikesh (home of kundalini yoga) says,

> *"Brahmacharya is a transmutation of the sexual energy, it is changing the whole energy from the sex centre to the higher centres. When it reaches the seventh centre samadhi happens. Learning and practicing the movement of vital energy is spiritual science…"*

Whereas Dr Carey comments on these teachings within bible metaphors,

> *"The metaphor of "the beast like a flying eagle" is a reference to the "seed-electrons" or "cell-nucleoli" (solar germs) in the organs of procreation becoming free to ascend. The elixir was the conserved essence of the procreative stream in the generative system."*

Both teachings describe the same process:

The procreative seed is processed in the ovaries and testes and the seminal fluid is processed in the prostate. If not wasted, these essences are reabsorbed by the body and becomes what Percival calls the "soil" in the spleen (the materials available for stem cell production).

Since the procreative seed is rich in electricity (specifically nitrogen and phosphorus), minerals, and nutrients, "saving" it improves the quality and volume of the "soil" in the spleen, this is what's known as the "offering up of animals (procreative germs)."

If the "soil" in the spleen *is* enriched by the reabsorbed procreative seed, the cells produced there will be remarkable also. Thus, we have the regeneration of the body occurring under divine law. An interesting constituent of procreative essences is spermine crystals. Spermine crystals are a nerve stimulant that facilitate cellular regeneration and support the helical structure of DNA/RNA.

Gnostics refer to the procreational seed, which becomes the furnishing product in the spleen as the **"salt of alchemy"** -- because all 12 cell-salts exist within it. Samael Aun Weor warns us not to spill it, but to transform it –

> *"because mastery is represented in the salt of the earth, which is in our sexual secretions."* **Samael Aun Weor 1954**

TESTES AND OVARIES –
The "Seed" Portion of Reproductive Essences

Spermatogonia are the least mature form of "sperm," and ovum are the female equivalent.

Simply stated, they are the stem cells. In their infant form they are pluripotent (still able to differentiate into a variety of different cell types).

If preserved, the sperm degenerate and their molecules are reabsorbed into the body – this is the true 10% tithe,

> *"The animal man actually robs his body of its cells (one tenth to be exact) in forming the germs of procreation. This is a fact which science will one day admit."*
> **Page 75, The Zodiac, and the Salts of Salvation by G W Carey**

> *"For within our seminal tract lie seeds of power and nobility possessing the nature of our Creator. These **seeds** are the conserved energy of the sun's radiance, being placed within our physical body where they can create or destroy. Restraining this force into a powerful energy will ennoble the student and evoke within his body the Sleeping Serpent (Kundalini). When this force leaves the body through an opening in the top of the skull man is no longer a prisoner in this illusion world but is united to his own central universe."*
>
> **Day Spring of Youth By "M"**

PROSTATE - The "Fluid" Portion of Reproductive Essences

The Prostate, or Kardia in philosophical writings is known as the Skene's gland in women -- ejaculatory tubes enter these glands.

An oily substance is excreted by the Prostate, which is subject to varied degrees of consistency, from a thin and volatile fluid to thick, fixed oil. This portion of sexual substance is alkaline in reaction, rich in calcium, phosphorus, nitrogen, lecithin, albumen, nucleoproteins, iron, and vitamin E – it is remarkably similar to the fluids of the nervous system (CSF, interstitial fluid etc.) -- theoretically, Life Force and vital fluids are the same thing existing in different states.

After seminal fluid is broken down in the prostate gland, and assuming it is not ejaculated it will make its way back through the capillaries into the blood.

Some experts consider the **prostate** to be "the seat of the Kundalini", others consider it to be the **coccygeal body** as we'll see later.

The Greek and Hebrew words for prostate, "Kardia" and "Kanda" respectively, both mean "Bulb." The word "prostate" comes from the word "proistamenoi" a Greek word meaning "priest" – priests lead the "**chi**ldren of God", and the prostate leads the **chi**/Life Force.

Professor Hotema states that, *"when the prostate and its "oil" act properly addictions can be dissolved."* The higher the quality and quantity of sexual vital essences, the more vibrant the Life Force sustaining the body.

My book, THE GOD DESIGN: Secrets of the Mind, Body and Soul explains more about the benefits of semen retention.

~ ~ ~

When Scripture talks about *"separating the sheep from the goats"*, it is highlighting the importance of leading with your head centre (Aries) and not letting carnal desire stemming from the sacral centre (Capricorn or Goat) lead you astray.

- CSF ventricles are shaped like a sheep or ram's head (Aries) with its curling horns.
- The procreative organs resemble a goat.

To summarise this part of the journey, the seed descended to the procreative organs where it can be preserved and reabsorbed into the body providing a powerful mineral base for the production of stem cells. The next location to visit is the kidneys.

KIDNEY – "NAPHTALI", "THE STRENGTH CENTRE"

The spleen is located on the left side of the body, we can assume that this is why Percival specifies the *left* kidney also. This image clearly illustrates the difference between left and right kidney connections.

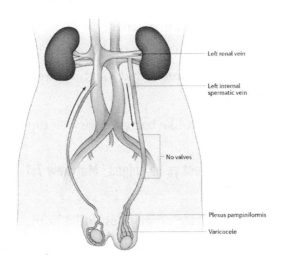

In **The Biology of Kundalini** page 43, Jana Dixon says, *"Tingles are felt especially on the left side of the body, and bubbles like champagne in the pelvis. Tingles and bubbles are always associated with increased kundalini flow."*

The metaphysical Bible dictionary tells us that the kidneys are symbolised by the biblical place called "Naphtali".

The kidneys are regarded as the body's most important reservoir of Chi (Life Force energy). The root word in "Kidney" is "kid", meaning "young goat" -- "kids" are "**children**". The inner child is the inner "chi" -- photon-light precipitating as nitrogen (fire) and phosphorus (light), of which there is a noted amount in the kidneys:

> *"And I won't tell you the amount of dynamite we could make with all the nitrogen and phosphorus we could extract from a kidney."* **Page 70, "The Doldrums, Christ and the Plantanism" By B R Garcia**

Professor Hotema explains that the bulk of Nitrogen and Phosphorus is stored in the **kidney organ-energy system**, which is probably why the kidneys are also known as the 'Root of Life'. The so-called, **kidney organ-energy system** includes, 1. the adrenal glands and 2. the testes or ovaries.

The adrenal glands sit like barrister wigs on top of the kidneys,

"Judge not, lest ye be judged" Matthew 7:1-3

The adrenals secrete a wide range of essential hormones that regulate metabolism, excretion, immunity, sexual potency, and fertility. Destruction of the adrenal cortex is fatal.

Criticism causes the secretion of detrimental hormones from the adrenals, so in a very literal sense the judgement of others causes our own self harm. The adrenals secrete a wide range of essential hormones that regulate metabolism, excretion, immunity, sexual potency, and fertility.

The kidneys control the balance of all bodily fluids and regulate the body's acid-alkaline pH balance, and we already know how important pH balance is toward the great regeneration due the bipolar, acid-alkaline, luni-solar nature of EVERY CELL IN THE BODY.

Poor memory, mind fog, fear, paranoia, and backache are all regarded as indicators of impaired kidney function and deficient kidney energy. Healthy kidneys drive courage and willpower.

Oxytocin (OT) is secreted by the pituitary gland and synthesized in the heart -- heart-derived oxytocin regulates kidney secretions and invigorates cardiac stem cell generation. Kidney phosphate secretions also rely on hormones released by the pituitary gland.

> *"We were struck by the fact that after removal of the pituitary body, the kidney may lose in the following few hours its power of secreting inorganic phosphorus."*
> **The Secretion of Inorganic Phosphate by the Kidney, L. Brull and F. Eichholtz**

Our next stop is the pancreas as this is where insulin is secreted.

PANCREAS – "ALL CREATOR"

The word pancreas is composed of the two roots "pan" meaning "All" (as in the all-pervasive source of creation) and "creas" meaning "to form". Biblical "bread" also symbolises the "All" -- *"Give us this day our daily bread."* The pancreas is referred to as the "bread pan," "**pan**try" or "**pan**nier" (breadbasket). The pancreas is innervated by the solar plexus (Bethlehem -- house of bread)

The pancreas is part of the endocrine and digestive systems. It is situated in the stomach, behind the navel and **secretes insulin.** Insulin doesn't just control the amount of blood sugar that goes into the muscles; **it also communicates with the genes.**

The key thing to know about insulin, is that it vitalizes DNA gene expression.

GENE EXPRESSION

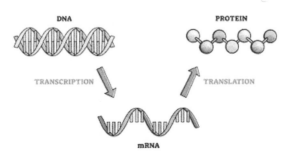

HEART CENTRE – "JERUSALEM," "ANAHATA"

Jerusalem symbolises *"the great nerve centre, just back of the heart."*
Metaphysical Dictionary by C Fillmore. It is the Anahata-cardiac
plexus, formed by nerves from the cervical ganglia.

"Jerusalem" is the Holy City; it represents the love centre in
consciousness. The loves and hates of the mind are precipitated
to this ganglionic receptacle of thought and are crystallized there.
THIS IS REALLY IMPORTANT BECAUSE LOVE STIM-
ULATES OXYTOCIN (OT) RELEASE.

The flow of the pituitary chemical oxytocin is enhanced by feelings of love and pure intention in the heart. Subsequently, CSF flow
increases and pressurizes so that the pineal can upgrade melatonin.
Melatonin upgrades will be explained more thoroughly later.

The endocrine thymus gland is associated with the cardiac plexus,
it is situated in the chest and begins to shrink when the genital
(goat) organs develop.

The majestic thymus also stocks the body with lymphocytes!
White blood cells, which are integral to the great regeneration as
described earlier in the spleen section.

> "Lymphocytes have the ability to travel freely through the body and are more abundant than any other of the body's wandering cells."
>
> **Page 50, The Thymus Gland by M Burnet**

Sir Burnet also states that the function of the thymus is deeply entwined with the information-carrying role of DNA. Deoxyribonucleic ACID is located in every cell nucleus, with the "solar germ" nucleoli explained earlier, and of course its messages affect our entire being! Therefore, the vibrations of the thymus have the ability to drive health *or* dis-ease. This ties in beautifully with Charles Fillmore's summary of Jerusalem,

> "What we love and what we hate here build cells of joy or of pain. In divine order it should be the abode of the good and the pure, but because of error concepts of the mind it has become the habitation of wickedness."
>
> **Charles Fillmore, The Metaphysical Bible Dictionary**

Professor Hotema says, *"We do not love with all our heart, but with all our Thymus Gland."*

In summary, feelings of love originating in the thymus cause the pituitary to secrete oxytocin and consequently the pineal upgrades melatonin.

The seeds journey then continues upward from the thymus to the thyroid, which is also innervated by cervical ganglia.

THYROID / THROAT / CERVICAL VERTEBRA 4 (C4) – "THE BAPTISM"

> *"Jesus was baptised of John in the fluids, the Christ Substance of the spinal cord."*
> **Page 46, The Tree of Life, by G. W. Carey**

Dr Carey clearly explains that broadly speaking, "John" is a metaphor for the "soul" which the masters and modern scientists agree is basically brain and nerve fluid.

> *"The soul swims in CSF"* **Randolph Stone.**

Dr Carey says that the word "John" is a chemical formula. Let's unpack this...

"Jesus" is said to be 30 years old when "John" baptises him. Jesus's 33 years of life symbolise the 33 vertebrae in the human spine, therefore we know that Jesus's 30th year must correspond with the 30th vertebrae of the spine. The 30th vertebra of the spine is C4, the 4th cervical vertebra.

The top of the thyroid gland is level with C4.

The Thyroid uses iodine (an electron donor) to produce thyroxine -- a powerful hormone which disinfects all the channels of the autonomic nervous system while we sleep! Without this biological iodine we would not be able to live.

In other words, thyroxine literally baptises (purifies) the body, mind, and soul... AND... the letters which spell John (previously Iohn) are found in the chemical formula of thyroxine.

Thyroxine

	Invisible, Creative, Astral Substance	Explanation
J/I	IODINE	There was no letter "J" in the Hebrew language, historically the name "John" would have begun with an "I" or a "Y".
O	OXYGEN	Oxygen corresponds with the Hebrew letter "Aleph", the "catalyst of creation".
H	HYDROGEN	Hydrogen corresponds with the Hebrew letter "Mem", the "water of consciousness".
N	NITROGEN	Nitrogen corresponds with the Hebrew letter "Shin", the "fire of life".

In summary, the Jesus seed is baptised by John (thyroid secretions). This leads us to the crucifixion.

DOUBLE CROSS – "BETHANY THE PLACE OF CROSSINGS," "CRUCIFIXION"

3 years after his baptism at C4, Jesus is crucified at age 33. Vertebral level 33 is at the medulla, near the entrance to the cerebellum, where the **double cross of crucifixion (transmutation)** is situated.

> *"Golgotha is the base of the human skull, where the spinal cord enters the brain. At this point occurs a **double nerve crossing**, made by Ida and Pingala and the vagus nerve."*
> **Page 144, God Man: The Word Made Flesh.**

it is formed by:

1. The Ida and Pingala (carrying the pituitary and pineal potencies)
2. The left and right vagus nerves (breath channels)

The crucifixion can actually be likened to an invigoration of potential...

> *"To crucify, means to add or to increase a thousand-fold. When electric wires are crossed, they set on fire all inflammable substances near them. When the Christed Seed crossed the nerve at Golgotha, the veil of the temple fell, and the*

> generative cells of the body were quickened or regenerated"
> **Page 65, God Man: The Word Made Flesh by GW Carey**

Dr Carey states that Moses' "swaddling clothes" are the mineral salts that protect or enrich the pineal and pituitary potencies on their DESCENT. On the seeds ASCENT, the "swaddling clothes" or cell-casing is removed at the double cross.

> **"The oil in the seed,** when born, is covered by a crust of mineral salts, which, when baptized in Jordan by John, is loosened in order that the shell may fall apart when the seed, goes over the cross, in order that the precious material may ascend into the pineal gland."
> **Page 62, God Man: The Word Made Flesh by G W Carey**

After its "crucifixion" the seeds constituents it enters the cerebellum which is the "tomb."

CEREBELLUM – "SPRIG OF ACACIA", "TOMB"

The cerebellum is synonymous with the Masonic "Sprig of Acacia" and the biblical place "Jericho." In fact, many esoteric symbols correspond with the sacred cerebellum:

- Dr Carey links the cerebellum with Taurus and Venus (Phosphorus) – *"Venus rules the awakened man."*
- Carey also proposes that the cerebellum is the "tomb" where "Jesus" was laid to rest.
- James Pryse refers to the cerebellum as "the magnetic, chemical centre" - **electrons are tiny magnets!**

The cerebellum is the "intuitive brain," which controls the involuntary functions of the body, such as breathing, circulation, sleeping, digestion and swallowing (autonomic functions).

The cerebellum deals with synergistic activity and makes everything work in harmony with everything else. It is the coordinating activity of the cerebellum that allows us to perform whatever function we choose. All the activity of the body – muscle tension, joint relaxation, hearing, vision, the relationship of every part of the body in time and space – sends impulses to the cerebellum.

The cerebellum is like an electrical loom -- electrical impulses give the cerebellum a perfect representation of the body's position in time and space.

> "The Creator dwells in you and his throne is the cerebellum. Prayers expressed by man in the cerebellum for righteousness are answered in the cerebrum, thus by prayer to God within, and in no other way can man overcome his adversary."
> **Page 140, God Man: The Word Made Flesh by G W Carey**

Imagine something you wish to manifest; this thought is pulled together from the neurons of the brain and an image is formed in the frontal lobe. If that image is held for long enough, a belief is created -- **and the "magnetic" cerebellum is activated.** Consequently, your actions are directed towards this newly established truth as an outcome.

> "The cerebellum clearly demonstrates the proper functioning of the intellect as it begins to move into Christ consciousness. As we become aware of the underlying activity that coordinates, balances, and harmonizes every action in creation, the cerebellum begins to receive this picture. Then we have available to us the information that represents the total body of creation, and we can become co-creators with the primary Creator."
> **Revelation the book of Unity by J. Sig Paulson and Ric Dickerson – Unity Magazine May 1975, Vol 155. No 5, Page 9**

The Cerebellum is the part of the brain that receives messages from the nervous system, it tells us what we think or feel about certain things. Eckhart Tolle explains that *"in the absence of awareness virtually all of your thoughts happen TO you instead of FOR you,"* meaning that the autonomous system is running riot and keeping you imprisoned to discordant thought patterns, thus you are not considered a "master of yourSELF".

When we give more power to the forebrain or rational part of our consciousness, we contribute to the atrophy and redundancy of the all-powerful cerebellum.

Consequences of the degenerate cerebellum:

- Unconsciousness during the transition between sleep and wake.
- **No recollection of astral experiences and difficulty remembering dreams.**
- Loss of reception between the body's higher and lower centers. Preventing the clarity of messages received from the Higher Self.

> "The objective mind being stilled, and the subjective mind in the attitude of receiving, draws to itself a new supply of Divine Essence. Life is renewed, electrified, strengthened, and purified. The other brain is the cerebellum, a negative organ or switchboard, switching the magnetic aura from the autonomic system to the cerebral and spinal nerves. **EACH BRAIN HAS ITS NERVOUS SYSTEM."**
> Revelation the book of Unity by J. Sig Paulson and Ric Dickerson – Unity Magazine May 1975, Vol 155. No 5, Page 9

This quote highlights the dual aspects of the nervous system: autonomic and central.

In summary, when the seed is crucified, it remains 2.5 days in the cerebellum which essentially switches, diffuses, or "resurrects" its energy from autonomic (involuntary power) to the central intelligence.

> *"...and on the third day ascends to the pineal gland that connects the cerebellum with the optic thalamus, the central eye in the throne of God."*
> **Page 90, God Man: The Word Made Flesh by G W Carey**

The cerebellum admits the "resurrected" current through the cerebellar lingula, into the fourth ventricle (CSF reservoir), where it is distributed into CSF, the pituitary and to the pineal via the posterior commissure.

This leads us to the next point on the journey, where we learn about the role of the optic thalamus.

OPTIC THALAMUS – "HOLY EYE"

> *"The positive pole, must be "lifted up" from the kingdom of earth, below the solar plexus, to the pineal gland which connects the cerebellum, with the optic thalamus."*
> **Page 24, The Zodiac and Salts of Salvation by G W Carey**

Specifically, **the "optic thalamus" is the basal part of the thalamus containing part of the optic nerve, chiasma, and optic tract.** In the 2nd century Galen defined the "optic thalamus" as two oval masses closely associated with the CSF ventricles on either side of the brain.

The two "oval masses" connected by the **Massa Intermedia** were designated the "optic thalami" because they were found to be involved with the processing and projections of visual reality.

The optic thalamus is also known as:

- Ophthalmos or optanomai – "The eyes of the mind"
- Thalamus Opticus – The Latin form of "Couche Optique"
- Optic Eminence – a mystic synonym
- "The Holy Eye", "The Eye of Providence", "The Eye Which Sleepeth Not"

- "The Eye which is the subsistence of all things." G W Carey

The image shows the optic thalamus as number 6. The front connects to the optic chiasma (above the pituitary) and the back connects to the epithalamus (including the pineal gland). This shows us what Dr Carey meant when he said that the optic thalamus connects the pituitary with the pineal.

> *"The optic thalamus, meaning "light of the chamber," is the inner or third eye, situated in the centre of the head. It connects the pineal gland and the pituitary body. The optic nerve starts from this "eye single."""*
>
> **Page 25, The Zodiac and Salts of Salvation by G W Carey**

Division

- 1 **Thalamus**
  - interconnection of afferent pathways to the cortex
- 2 **Metathalamus**
  - the dorsal part of the thalamus containing the nuclei of the visual and auditory tracts
- 3 **Hypothalamus**
  - a control centre of autonomic and endocrine functions
- 4 **Epithalamus**
  - its main part is the pineal gland
- 5 **Subthalamus**
  - nuclei and tracts involved in motor and emotional neuronal patterns
- 6 **Thalamus opticus**
  - the basal part containing a part of the optic nerve, chiasma, and optic tract

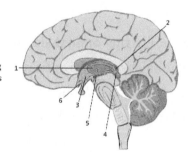

Sagittal section of the brain

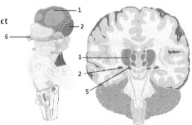

The Massa intermedia connects the two lobes of the optic thalamus, and in a way rules our very existence – we will journey there now.

MASSA INTERMEDIA
– "REALITY "PROJECTOR"

The Massa Intermedia is in the centre of the optic thalamus. Its surfaces form part of the lateral wall of the **third CSF ventricle.**

"Reality" literally projects from the massa Intermedia.

At the Massa Intermedia is the centre of the brain's torus field, this centre point is a plane of inertia built by the brain's resonance.

Our senses simply absorb and filter the frequencies that we project from the internal to the external back into consciousness and perceives them as "reality".

The frequencies flow to the optic thalamus, where they are directed to the microcosmic sun (pineal) and the microcosmic moon (pituitary), allowing us to "see" our "reality".

The frequencies vibrate the cerebrospinal fluid inside the 3rd ventricle, giving rise to the shapes and forms of so-called "physical life". The entire experience of life is generated in this way.

The sun and Moon, or the microcosmic pineal and pituitary work together as an all-natural electromagnetic generator. This natural mechanism is what enables our hearts to beat and all other organic processes to take place.

The generator charges our temple body and gives everything life and energy. CSF acts as a conductor because of its high salt level.

Having realised our creative power initialised in the massa inter-media and third CSF ventricle, we see that we can indeed write our own life stories. Subsequently neurons light up creating new and powerful pathways of thought and possibility in the cerebrum or "most high."

CEREBRUM - "MOST HIGH"

The cerebrum controls the central nervous system.

The root of cerebrum is "cere" meaning "wax". From the same root comes cerebral, or cere-al meaning "seed".

The cerebrum is synonymous with the Biblical character "Abram" or "Abraham". The cerebrum is "the upper brain", "most high" or "Jehovah". It is the place of "the crystalline dew".

Other hidden symbols such as the "Island of Reil", the "Island of Patmos" and the "North Pole" correspond with the central lobe of the cerebrum.

Dr Carey states that *"Aries (Cerebrum) and Taurus (Cerebellum) lay down the law on the other parts of the body."*

Before differentiation by the pituitary and pineal, the cerebral substance is neutral; it consists of all things.

> "The cerebrum is truly the fountain-head of life, and within it is precipitated the mysterious semi-fluidic and wax-like substance which is termed "grey matter." **The infinitesimal mineral particles** in this grey matter constitute the magnet, which attracts humid air and the Spirit of life into the brain

reservoir or claustrum."

Page 215, The Zodiac and Salts of Salvation by G W Carey

So, the refined essences refresh the contents of the cerebrum, and another cycle immediately begins at the claustrum which integrates the Light. But we can't conclude the journey without knowing what climaxes occur in the pituitary and pineal.

PITUITARY – "MOON"

The pituitary is the first port of call,

> "When the Spirit Fire is lifted up through the thirty-three segments of the spinal column, and enters into the human skull, it passes into the pituitary body (Isis), where it invokes Ra (the pineal gland) and demands the Sacred Name."
> **Page 194, The Secret Teachings of All Ages, Manly P Hall**

The pituitary is the "master gland" of the endocrine system, the release of pituitary secretions corresponds with the electromagnetic radiation and the gravitational pull of the moon. Macrocosmically, the moon regulates the oceans and tides in time with its monthly cycle. Microcosmically the pituitary hormones regulate and refresh our soul (fluid body) on a monthly cycle also, this is possible due to presence of magnetite crystals.

> "The pituitary gland is considered an important energetic organ, in that it contains magnetite, a magnetically sensitive

> compound." **Page 63, The Complete Book of Chakra Healing: Activate the Transformative Power by Cyndi Dale**

Posterior Pituitary Lobe:

The "lunar" portion of the regenerative seed is said to be a posterior pituitary lobe secretion. "Pituitrin" is the historical name for combined posterior lobe secretions. Namely, oxytocin (OT) and vasopressin (VP).

Oxytocin (OT):

> *"The moment oxytocin levels go up the brain's survival centres cool off. The amygdala **(an anagram for Magdalene)** slows the circuits of fear, sadness, pain, anxiety, aggression, and anger. Then the only thing we feel is a love for life."*
> **The Reason Why Kindness Makes Us Happy, Unlimited Blog by J Dispenza (brackets my own)**

Fear, sadness, pain, anxiety, aggression, and anger are all emotions that counteract the great regeneration – oxytocin is a powerful antidote to these debilitating states of consciousness.

The chemical composition of oxytocin is similar to that of protoplasm (lunar germ). Oxytocin has many important roles in the temple body, including the stimulation, production, and mitosis of **stem cells!**

Lack of oxytocin brings on premature aging, but increasing oxytocin levels through meditation, compassion and forgiveness actually slows the aging process by improving the behaviour of stem cells.

- Oxytocin is known as the "love" hormone
- Christ is Eros (the love God)
- Percival refers to the lunar germ as the "Christ Lunar Germ"

Therefore, a correlation between what experts historically referred to as "protoplasm – the lunar germ" and what is now known as oxytocin definitely exists.

Vasopressin (VP):

Vasopressin mediates stress and stabilises circulation. Vasopressin has the monumental task of maintaining the appropriate volume of water in the space that surrounds cells of the body, the extracellular matrix. This allows proper cellular function. Vasopressin also helps the body maintain fluid volume.

In other words, vasopressin is indispensable to the vital fluids (lunar aspect) of the body.

Pituitary secretions act as a catalyst for pineal activity thus producing the felt sensation of the great regeneration.

> "In the brain it first activates the pituitary, the feminine, nega-
> tive pole, causing it to send a stream of **bluish** solar electricity
> thru the infundibulum to the pineal, the male, positive pole,
> thus completing the circuit."
> **Page 522, The Son of Perfection (Part 1) by H Hotema**

PINEAL – "SUN", "PINNACLE"

The pinnacle of enlightenment produced in the pineal gland has two major aspects.

1. The stimulation of nitric oxide release (the kundalini)
2. The upgrade of melatonin (the nectar of the Gods)

These two operations are synchronous and will now be explained.

The pineal gland is bathed in highly charged cerebrospinal fluid (CSF). The pineal contains calcite crystals that are piezoelectric, endothelial cells that generate nitric oxide (NO) and pinealocytes that mediate nitric oxide release.

The *"stream of blue solar electricity"* travelling from the pituitary to the pineal, mentioned by Professor Hotema is nitric oxide (the sky is blue due to the presence of nitrogen).

When we are in a state of harmony the posterior pituitary lobe secretes more oxytocin and vasopressin, causing pineal crystals to vibrate more rapidly, this increases our vibratory frequency (centred in the mass intermedia).

This is the alchemical wedding; lunar and solar body's uniting in a climax of power and healing,

> *"Jesus said, if these two make peace with each other in this one house, they will say to the mountain, "Move Away," and it will move away."* **Thomas Logia 48**

The harmonic state initiated by pituitary secretions causes the rapid release of nitric oxide and stimulates the pineal to upgrade melatonin into DMT and a host of other widely beneficial neurochemicals.

We'll look at the incredible benefits of increased nitric oxide flow now, before moving on to appreciate the role of DMT and its benefactors.

NITROGEN - The "Spirit Fire"

NO (Nitric Oxide) is basically nitrogen + oxygen, it is a colourless gas that is **formed by the oxidation of nitrogen.**

Nitric oxide influences pineal metabolism (DMT synthesis) and the blood flow of the gland.

When photon light is absorbed by the body, it forms Nitric oxide (NO). Nitric oxide is essential for the metabolism of ALL CELLULAR REGENERATION. Nitric oxide is a molecule of health – the more the better! The benefits of self-produced nitric oxide are prolific! For example:

- Nitric oxide is involved with everything from the binding and release of oxygen and haemoglobin, to inhibiting inflammation.
- Nitric oxide is even linked to the destruction of viruses, parasites and malignant cells in the airways and lungs.

In the great regeneration nitric oxide stimulates mRNA, the messenger of DNA. Nitric oxide is a free radical, this explains why the great regeneration process is about dissolving old structures (physically, mentally, and emotionally).

In **The Biology of Kundalini** page 43, Jana Dixon says, *"Tingles and bubbles are always associated with increased kundalini flow. There is some indication that the tingles are associated with increased nitric oxide."*

Remember, "air" is mostly nitrogen and oxygen. Electrons in the air are the "Life Force" in the **"breath of life."** More than we breathe for "air" we breathe to take in Life Force in the form of electrons. Nostril breathing increases nitric oxide in the body (hence all of the pranayama's that involve nasal breathing).

In Hinduism a common mantra is OM, Christianity altered OM and created their "AMEN." Chanting OM boosts the production of Nitric Oxide (NO) in the body. In fact, most spiritual practices boost nitric oxide flow and release excess carbon dioxide poison.

MELATONIN UPGRADES - "DMT" etc.

Like the sun, the pineal wakes us up with its serotonin secretion and puts us to sleep with its melatonin secretion.

Serotonin/Melatonin is the nectar of life, "sero" means seed, and "mellis" is the Greek word for honey. A glimpse at their chemical structures reveals the letters that spell honey.

Serotonin   Melatonin

Melatonin is a timekeeper for gonadal maturation, so the pineal is implicated as the trigger for puberty and the initiation of the great monthly regeneration cycle.

> *"The first seed is formed in the solar plexus of every individual, commencing at the age of puberty."*
>
> **Page 48, God Man: The Word Made Flesh by G W Carey**

Melatonin and serotonin increase DNA synthesis and induce mitosis (cell renewal) -- a small electrical signal is sent through the double helix of DNA which instigates a signal allowing hydrogen bonds to unzip so that DNA can replicate.

Serotonin and melatonin are derived from "tryptophan."

IN FACT, ALL OF THE ENHANCED BIOCHEMICALS OF ENLIGHTENMENT ARE DERIVED FROM TRYPTO-PHAN. Looking at the tryptophan/DMT pathway really helps us understand the concept of melatonin upgrades.

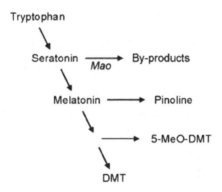

Tryptophan is literally a precursor to DMT.

DMT is known as the "spirit molecule" – an endogenous hallucinogen which is the source of visionary light in transpersonal or "mystical" experiences.

> "DMT in the pineal glands of biblical prophets gave God to humanity and let humans perceive parallel universes"
>
> **Page 140, Conversations on the Edge of the Apocalypse by D J Brown**

When pituitary hormones stimulate the pineal to upgrade melatonin into DMT, the DMT stimulates photon-light emissions from DNA, therefore we actually "shine brighter."

But DMT does not and cannot work alone in the great regeneration! **A whole host of other blissful biochemicals are produced too.**

The hallucinogen Pinoline is a major contributor to the felt experience of enlightenment... the subconscious mind uses it to communicate with the deeper spheres allowing prophetic visions to be seen on the timeline. Lack of sleep inhibits pinoline production.

> "When the kundalini hits the pineal gland it ionizes the spin ratio of serotonin, its electrons interchange altering its chemical nature. The molecule itself is reconfigured to its highest potential – pinoline."
>
> **Page 255, A Beginner's Guide to Creating Reality by Ramtha**

DMT, Pinoline, and ALL of the biochemicals of enlightenment are detailed in my book THE GOD DESIGN: Secrets of the Mind, Body, and Soul.

The biochemicals of enlightenment travel from the pineal to the optic thalamus via the peduncles (two strands of nerve fibres). The optic thalamus, along with the 3rd ventricle receive the upgraded biochemicals, which has a "foaming affect" in CSF, causing it to rise and multiply.

When the supply of cerebrospinal fluid exceeds the volume of the central canal, and the ventricles of the brain it seeps over and bathes the nerves, resulting in the experience of an intense physical and spiritual bliss.

> "Like yeast, the seed that comes fourth from the pineal, **expands and causes the oil in the spinal cord to multiply."**
> Page 97, God Man: The Word Made Flesh by G W Carey

"Flashes of light" and the ultimate clarity of mind occurs, the massa intermedia is freed from discordant cycles, limitless potential is realised, and the great regeneration is underway.

TERMINAL FILAMENT – "BRIDGE TO IMMORTALITY"

One last thing to consider is what's known as the "terminal filament" also known as the "filum terminale."

In **Thinking and Destiny,** Harold Percival says that the terminal filament is atrophied or "clogged up" in adults. He says it takes 13 lunar cycles to reopen or regenerate the terminal filament. Assuming the procreative seed has been preserved and raised.

The spinal cord reaches from the base of the brain and becomes narrower at the centre of the body (See T12). The image shows the terminal filament extending down from the spinal cord, through the cauda equina, all the way down to the first segment of the coccyx, in the vicinity of the "coccygeal body."

The terminal filament is esoterically known as a "bridge" or passage from the autonomic nervous system to the central nervous system.

> "After the 13th cycle, the seed still descends in the right sympathetic nerve, but instead of ascending in the left sympathetic nerve to the kidney, it connects with the central nervous system via the "bridge" (terminal filament)."
>
> **Page 722, Thinking and Destiny by Harold W Percival**

Since the terminal filament terminates in the coccygeal body, this will be our last stop on the journey of the great regeneration.

COCCYGEAL BODY (GLOMUS COCCYGEUM – KUNDALINI GLAND):

According to Jana Dixon, the "Kundalini gland" is the **coccygeal body** and not the prostate (discussed earlier).

The coccygeal body is a gland situated at the bottom of the coccyx and is said to "pulsate" when kundalini (nitric oxide) flow increases, this has been esoterically described as in "inward ejaculation up the spine."

Carey's description of the kundalini rising from this point also sounds like nitric oxide gas.

> *"That which is within this canal is of a substance more like steam or gas than anything else."*
> **Page 166, God Man: The Word Made Flesh by G W Carey**

The coccygeal body is comprised of **epithelioid cells**, known to generate nitric oxide (NO). It is a nexus of the body-mind systems: hormonal, blood, sympathetic and parasympathetic nerves, and the immune (lymph) system. Therefore, it is clearly a key to the homeostasis and regeneration of the body.

CONCLUSION

There is a perpetual cycle occurring in the temple body. The cycle can cause degeneration or regeneration physically, mentally, and spiritually depending on our vibration and choices.

The regeneration of the fluidic (lunar) body happens monthly, coinciding with the moon.

The regeneration of the mineral (solar) body happens yearly, coinciding with the sun.

Light in the form of photons (electromagnetic energy) is received by the brain and differentiated by the pineal and pituitary.

The two potencies flow through the autonomic nervous system, through the semilunar ganglion and into the solar plexus where they merge and conceive the seed in the spleen.

After conception some of the seed will automatically flow up the spinal cord from T12, Percival calls this "automatic reclaiming".

The remainder enters the vagus nerve and descends to the procreative organs where it is further vivified (if not expelled).

Through what Percival calls "voluntary reclaiming", the vivified seed is then reabsorbed into the body where it begins its ascension to C4 for the baptism.

On its path to C4 it travels through other vital organs including the kidneys and the heart. Love in the heart stimulates oxytocin release by the pituitary, this chemical is the catalyst for pineal metabolism.

After its baptism by thyroxine the seed arrives at the double cross of Ida and Pingala and the vagus nerve where it is crucified.

After crucifixion the seed is sent to the tomb (cerebellum).

The cerebellum admits the "resurrected" seed through the cerebellar lingula, into the fourth ventricle (CSF reservoir), where it is distributed into CSF, the pituitary and the pineal via the posterior commissure.

This causes the pituitary to secrete floods of oxytocin and vasopressin which consequently invigorates the pineal.

The stimulated pineal then glows rich with nitric oxide and upgrades melatonin to **DMT and the other biochemicals of super consciousness (Read "The God Design" for details).**

The enhanced CSF is "Amrita -- the nectar of the Gods" -- magnetized and charged (**ionized).

Thanks to the spirit-fire catalyst (nitric oxide) it is multiplied and flooded with DMT, pinoline, and the other biochemicals of super consciousness.

Immediately after, the process will begin again... and again...

Each cycle permits some of the essence to descend the central CSF canal via the ventricles, thus the temple body is purified by degree.

After the thirteenth round the hollow through the terminal filament is said to be fully cleared, thus the seed can travel directly from the coccygeal body to the brain.

The biochemicals of the great regeneration (sacred secretion) enhance consciousness, cognitive abilities, and health along all lines.

An added benefit is the reinvigoration of the long-term memory powered by the hippocampus; this is known as "accessing the book of life (DNA blueprint)" -- it's the recollection of the true self.

The Hippocampus is a primary region for neurogenesis and contains multipotent neural stem cells.

This entire process is affected by two major INPUTS!

The input of thought and emotion via self and other affects our vibratory frequency and in turn our lymphatic systems. The outcome will be reflected by our body and circumstance. Low frequencies diminish our ability to regenerate.

The input of food and drink (saliva) affects the bodies pH – acidic pH diminishes our ability to regenerate.

As stated earlier, **there is a perpetual exchange occurring between blood, CSF (nerve/interstitial fluid), sexual vital fluids and lymph.**

All of these vital fluids are affected by saliva (dietary choices). Hence Carey's claim that saliva can bring salvation. Acidity is a lack of electrons (life force).

> "A river (saliva) went out of Eden to water the garden, and from thence it was parted and became four heads. Pishon is the urine; Gihon is the intestinal tract; Hiddekel is the blood; Euphrates the nerve fluids, especially the creative."
> **Page 56, The Tree of Life by G W Carey**

IN OTHER WORDS, SALIVA IS THE BASE SUBSTANCE FROM WHICH ALL OTHER BODILY FLUIDS ARE FORMED.

Since saliva is paramount to the great regeneration, it stands to reason that what we eat, and drink is important also. In "The God Design: Secrets of the Mind, Body and Soul" there is an entire chapter called "How to Raise the Sacred Secretion," it gives in depth advise on how to awaken your own kundalini and upgrade melatonin. However, I will give a summary of guidance to conclude this book also.

The Simple Seven:

1. **Sun**

    Make sure you receive daily exposure to NATURAL light; photon energy is imperative to the health of the cellular terrain.

2. **Air**

    Take time to breathe deeply; even five minutes of deep nostril breathing each day will improve your health along all lines.

3. **Sleep**

    Thyroxine clears the lymph system during sleep, 7-8 hours of good rest will allow the parasympathetic nervous system to clear toxins and useless thought debris from the day. Stillness and peace can be induced by silently repeating "I am Light" and imagining the light within you increasing.

4. **Exercise**

    Your body was made to move, it likes to move... movement creates energy; dance, walk, do yoga, run, cartwheel whatever you like – just make sure you get your heart rate up for a minimum of 20 minutes daily.

5. **Water**

Hydration is everything, so drink plenty of good quality water. Watch out for pH levels in bottled water, try to drink water that is pH 7 or above.

6. **Food**

Avoid synthetic, processed foods wherever possible. The Daniel fast is the optimum eating regime.

7. **Preservation**

Each month, when the moon is in your sun sign refrain from climax induced by any sort of sexual stimulation. You can find these dates on an App such as Deluxe Moon – please observe the tropical and sidereal dates to ensure full preservation (there is a video about this on my YouTube channel).

SOURCES AND BIBLIOGRAPHY

BIBLES:

"The King James Bible Version (KJV)"
"The Besorah Of Yahusha Natsarim Bible Version (BYNV)"
"The New International Bible Version (NIV)"
"The Message Bible Version (MSG)"

BOOKS (Alphabetised by surname):

"Healing Mantras" Thomas Ashley-Farrand
"Vol. 1, The anatomy of the human body" C Bell & J Bell
"The Secret Doctrine" Vol 1. Madame Helena P. Blavatsky
"Isis Unveiled: The Secret of The Ancient Wisdom Tradition" Madame P. Blavatsky
"The Kundalini Process" Wim Borsboom
"Conversations on the Edge of the Apocalypse" D J Brown
"The Secretion of Inorganic Phosphate" L. Brull and F. Eichholtz
"The Light of Egypt" Thomas H. Burgoyne
"The Science of The Soul and The Stars" Thomas H. Burgoyne
"The Thymus Gland" M Burnet
"God-Man: The Word Made Flesh" George W. Carey and Ines Eudora Perry
"Relation of The Mineral Salts of The Body to the Signs of The Zodiac" George W. Carey
"The Tree of Life" George W. Carey
"Dictionary of Symbols" J C Cirlot

"Philosophical Transactions Series B, Biological Sciences" Francis Crick
and Christof Koch
"The Complete Book of Chakra Healing" Cyndi Dale
"The Twelve Powers of Man" John Fillmore
"Metaphysical Bible Dictionary" Charles Fillmore
"Talks on Truth" Charles Fillmore
"The Doldrums, Christ and the Plantanism" B R Garcia
"The Occult Anatomy of Man" Manly P. Hall
"The Secret Teachings of All Ages" Manly P. Hall
"The Facts of Nutrition" Hilton Hotema
"The Son of Perfection" Hilton Hotema
"The Secret Doctrine of the Rosicrucian's" Magus Incognito
"Light on Yoga" B.K.S. Iyengar
"The Biology of Kundalini" Justin Kerr
"The Perfect Way" Anna Kingsford
"The Gospel of Thomas: The Gnostic Wisdom of Jesus" Jean-Yves Leloup
"Day Spring of Youth" M
"The Cosmic Serpent" Jeremy Narby
"Revelation the book of Unity" J. Sig Paulson & Ric Dickerson
"Thinking and Destiny" Harold W. Percival
"The Living Message" Eugene H. Peterson
"A Beginners Guide to Creating Reality" Ramtha
"Endogenous Light Nexus Theory of Consciousness" Karl Simanonok
"The Essenes, the Scrolls, and the Dead Sea" Joan E. Taylor
"Vallalar's Vision of Nuclear Physics" T Thulasiram
"The Perfect Matrimony" Samael Aun Weor
"Practical Magic" Samael Aun Weor
"Esoteric Science Vol 1" J S Williams

ONLINE SOURCES:

The Metaphysical Dictionary at www.truthunity.com
Electric Science Magazine
Strong's concordance at www.biblehub.com
www.collinsdictionary.com
www.neuroquantology.com
www.researchgate.net

BOOKS AND PLATFORMS CREATED BY THE AUTHOR KELLY-MARIE KERR

BOOKS:

- **Christ Within, Heaven on Earth**
  A concise description of the journey of the glorious sacred secretion (transcript of True anointing YouTube video).

- **The God Design, Secrets of the Body, Mind and Soul**
  A thorough study and explanation of both the spiritual and physical elements that form the phenomena known as the sacred secretion. Including the full details of the biochemicals of enlightenment.

- **Elevation, The Divine Power of the Human Body**
  The Bible book of Revelation explains the true science of enlightenment: body, mind, and soul in a dramatic, fantastical, and epic parable only 22 chapters. Elevation debunks the symbols and myths providing truth and clarity to its reader.

- **The Cell of Life, Awakening and Regenerating**
  A full disclosure of the *3-Fold Enlightenment* or Great Regeneration, revealing the scientific parallel of the "Jesus" seed born in the body every lunar month. The "seed" is our opportunity for TOTAL renewal and regeneration.

> *"Every 29.5 days a seed is born in, or out of the solar plexus - the oil unites with the mineral salts and thus produces the monthly seed which goes into the vagus."*
> **Page 90, GOD MAN: The Word Made Flesh by G W Carey**

- **Super Soul Calendar, Keep Track of your Regeneration Seasons**
  A full year calendar providing the sidereal and tropical dates for the moon entering each star sign (zodiac), plus guidelines and meditation techniques to help you on your journey.

PLATFORMS:

- Youtube channel, "Kelly-Marie Kerr"
- Website, www.seekvision.co.uk
- TikTok, @seekvision
- Instagram, @seekvision
- Facebook, @seekvision33
- Twitter, @seekvision33
- Patreon, Seek Vision (Kelly-Marie Kerr)

APP:

- **Freedom Yoga**

  Both free and pro versions of **Freedom Yoga** are available for **android and iOS**. This App was created to allow you to customise the length and focus of your yoga practice. Each practice includes mantra (Scripture), prayer (meditation/manifestation) and physical postures.

  You may need to type "Seek Vision Freedom Yoga" to successfully find the app in stores.

Made in the USA
Middletown, DE
10 March 2024

51178780R00070